THE WORLD'S SMARTEST ANIMALS

Dolphins

by Joanne Mattern

BLASTOFF! READERS 3

BELLWETHER MEDIA • MINNEAPOLIS, MN

Blastoff! Readers are carefully developed by literacy experts to build reading stamina and move students toward fluency by combining standards-based content with developmentally appropriate text.

Level 1 provides the most support through repetition of high-frequency words, light text, predictable sentence patterns, and strong visual support.

Level 2 offers early readers a bit more challenge through varied sentences, increased text load, and text-supportive special features.

Level 3 advances early-fluent readers toward fluency through increased text load, less reliance on photos, advancing concepts, longer sentences, and more complex special features.

★ Blastoff! Universe

This edition first published in 2021 by Bellwether Media, Inc.

Library of Congress Cataloging-in-Publication Data

Names: Mattern, Joanne, 1963- author.
Title: Dolphins / by Joanne Mattern.
Description: Minneapolis, MN : Bellwether Media, 2021. | Series: Blastoff! readers: the world's smartest animals | Includes bibliographical references and index. | Audience: Ages 5-8 | Audience: Grades 2-3 | Summary: "Simple text and full-color photography introduce beginning readers to dolphins. Developed by literacy experts for students in kindergarten through third grade"-- Provided by publisher.
Identifiers: LCCN 2019059254 (print) | LCCN 2019059255 (ebook) | ISBN 9781644872406 (library binding) | ISBN 9781618919984 (ebook)
Subjects: LCSH: Dolphins--Behavior--Juvenile literature. | Animal intelligence--Juvenile literature.
Classification: LCC QL737.C432 M38 2021 (print) | LCC QL737.C432 (ebook) | DDC 599.5315/13--dc23
LC record available at https://lccn.loc.gov/2019059254
LC ebook record available at https://lccn.loc.gov/2019059255

Editor: Betsy Rathburn Designer: Jeffrey Kollock

Printed in the United States of America, North Mankato, MN.

Table of Contents

Smart Swimmers

Dolphins are **mammals**. They are part of the **toothed whale** family. They live in **pods** of up to 20 dolphins.

These animals are very playful. They love to jump and spin with their friends!

Dolphins live in oceans all over the world. Some even swim in rivers!

These animals use their **intelligence** to survive. They **communicate** with their friends. They even use tools!

Talk and Tools

Dolphins use their intelligence in many ways. They make sounds to communicate.

Some sounds warn of danger. Others tell when it is time to play. **Echolocation** helps dolphins communicate over great distances!

These smart animals use tools. Some carry sponges as they search for food. This protects their **snouts** from rough rocks.

Others use shells to scoop up fish. They eat the fish inside the shells!

snout

Dolphins can learn new things, too. They learn by watching other dolphins.

mother with calf

Calves learn new games by watching other dolphins play. They watch their mothers to learn how to **forage**!

Dolphins love to play! This is another sign of intelligence.

They make toys out of objects in the water. They chase each other and play tag. They even play with bubbles!

Learning About Dolphins

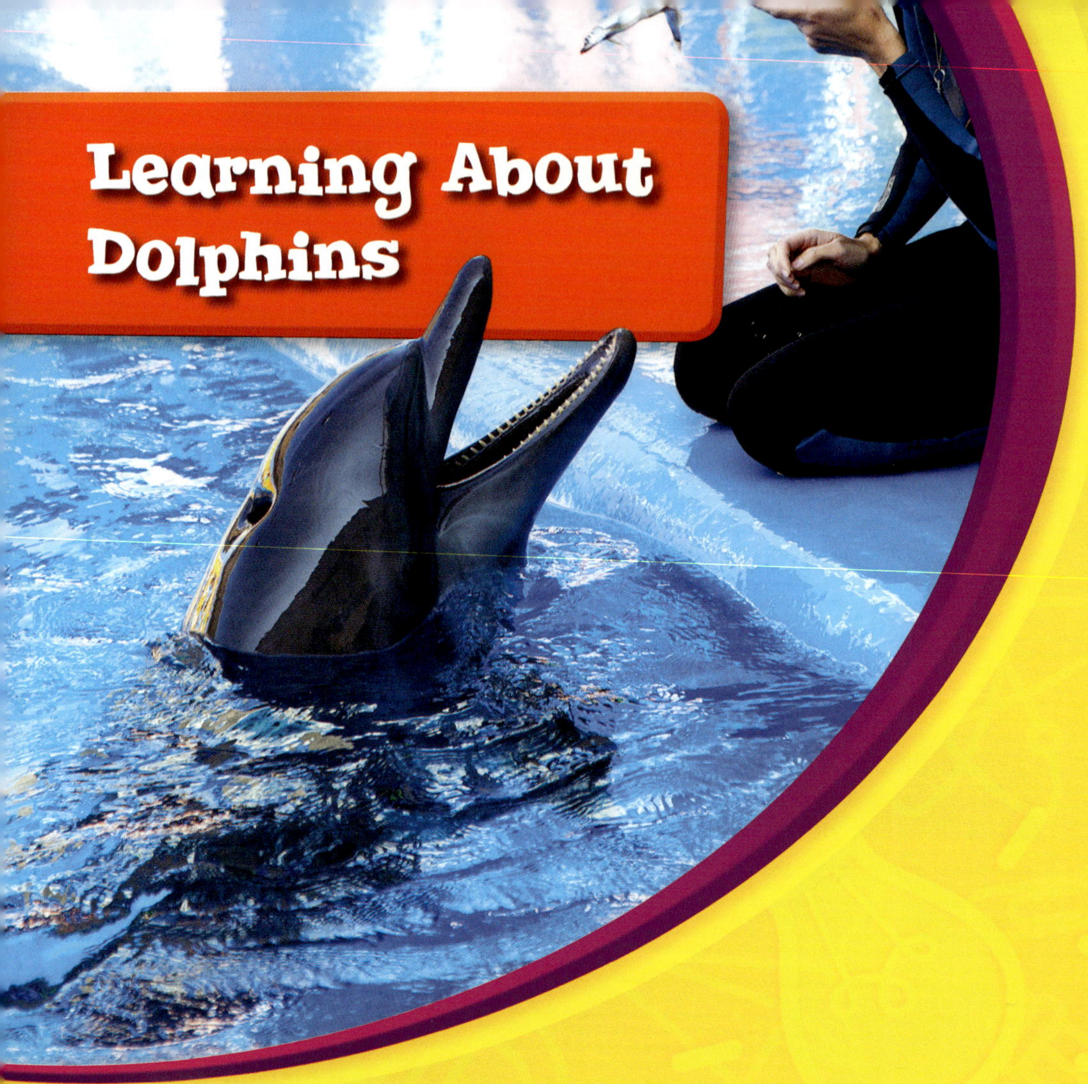

Many scientists study dolphins. They want to learn how dolphins communicate. Some scientists have taught dolphins **sign language**. Others teach dolphins to ask for toys and food using touch screens!

Dolphin Study

Question

Can dolphins work together to solve problems?

Process

1. Dolphins were trained to press buttons.
2. An area with two buttons was set up.
3. Dolphins were told to press the buttons.
4. When dolphins pressed the buttons at the same time, they got a fish.

What Happened?

- Dolphins did not get fish unless they pressed the buttons at the same time.
- The dolphins learned to press the button at the same time.

Answer

- Dolphins are able to work together to solve problems.

Some scientists do **experiments** with dolphins. They find out how dolphins think and learn.

Amazing Dolphin

Name

- Kelly

Species

- bottlenose dolphin

Famous For

- Learned to collect pieces of paper to get fish from trainers
- Tore paper into small pieces to get even more fish
- Rewarded with a lot of fish for catching a gull
- Used a fish to attract other gulls, earning even more fish from trainers

Other scientists train dolphins. Some dolphins work for the United States Navy. They find bombs underwater!

Scientists study dolphins to learn more about these smart animals. The studies teach scientists more about human brains.

Dolphins are not just playful swimmers. They are also one of the world's smartest animals!

Glossary

calves—baby dolphins

communicate—to share thoughts and feelings using sounds, faces, and actions

echolocation—the use of sound waves and echoes to determine where objects are; some dolphins also use echolocation to communicate.

experiments—tests used to try new ideas or answer questions

forage—to go out in search of food

intelligence—the ability to learn or understand

mammals—warm-blooded animals that have backbones and feed their young milk

pods—groups of dolphins that hunt together

sign language—a language made up of hand gestures

snouts—the long noses and mouths of some animals

toothed whale—related to a group of ocean mammals that have teeth; dolphins, porpoises, and orcas are all toothed whales.

To Learn More

AT THE LIBRARY

Pringle, Laurence. *Dolphins!* Honesdale, Pa.: Boyds Mills Press, 2019.

Schuh, Mari. *The Supersmart Dolphin.* Minneapolis, Minn.: Lerner Publications, 2019.

Swanson, Jennifer. *Absolute Expert: Dolphins.* Washington, D.C.: National Geographic, 2018.

ON THE WEB

FACTSURFER

Factsurfer.com gives you a safe, fun way to find more information.

1. Go to www.factsurfer.com.
2. Enter "dolphins" into the search box and click 🔍.
3. Select your book cover to see a list of related content.

Index

The images in this book are reproduced through the courtesy of: Andrea Izzotti, front cover, p. 1; Joost van Uffelen, p. 3; alekss-sp, pp. 4, 5; EpicStockMedia, p. 5; Helmut Corneli/ Alamy, pp. 6, 7; Tory Kallman, pp. 8, 9; VarnaK, p. 9; Yann hubert, p. 10; Willyam Bradberry, pp. 10, 11, 14, 15, 21; Chris Allan, pp. 12, 13; Elena Larina, pp. 13, 16; toshiharu_arakawa, p. 13; Darios, pp. 14, 18; fxphotovenezia, p. 17; Wikimedia Commons, pp. 18, 19; skvalval, pp. 20, 21; WitthayaP, p. 22.